GOING WITHOUT THE FLOW:

Prostrate with a prostate

I0391007

Adam Yamey, who was born in London, is a dentist and an author. In addition to two novels set in 19th century South Africa ("Aliwal" and "Rogue of Rouxville"), he has written books about the Balkans ("Scrabble with Slivovitz", "Albania on my Mind", "Rediscovering Albania", and "From Albania to Sicily") and South Africa ("Exodus to Africa" and "Soap to Senate: a German Jew at the dawn of apartheid"). He lives in London with his wife and daughter.

Published by Adam R Yamey

[A lulu.com project]

ISBN: 978-1-326-89494-8

This work is also available as a Kindle version
www.adamyamey.com

Going

Without

the flow

Prostrate with a prostate

Adam Yamey

Dedicated to the caring team at Holland Park Surgery

Manneken Pis statue in Brussels (Belgium)

[Source: Wikipedia. Photographer: "Myrabella"]

There is always an irrational aspect of fear. This increases in proportion to the amount of time which one has available to anticipate the potentially fearful event. If, out of the blue, a tile begins falling from a roof above you, you might have only a fraction of a second (or even no time) to worry about it hitting you. On the other hand, if you have a difficult task (e.g. an examination or a job interview) planned days or months ahead, then you have plenty of time to become increasingly anxious about it.

I have always been filled with fear at the prospect of any physical intervention on my body. This may come as a surprise to anyone who knows that I am a dentist, who makes a livelihood from trying to assist people who fear my interventions, but this is the case.

For example, from an early age, I have feared going to the barber, an experience that most people enjoy. I am not certain whether this fear of having my hair cut originated from hearing the tale of the barber of Fleet Street, who used to deliver his unsuspecting customers to the basement of the butcher next door, in order for them to be turned into sausage meat. I still cling onto the armrests of the barber's chair, just in case… Or, did my fear arise from the worry that I might be injured or infected by the scissors or the cut-throat razors, which are still used today?

There is yet another possible source of my 'pre-barber angst'. This dates back to the 1950s, when I was less than ten years old. In those days, I used to be taken to a large hairdressing salon in Golders Green Road, where Mr Pearce attended to my *coiffure*. The salon was filled with a nauseous odour, that of people having the split ends of their hair singed with the flame of a lighted taper. What, I wondered, would have happened had Mr Pearce begun to singe my hair? Would my head have erupted into a fiery ball? Well, this never happened. My beloved, but neurotic, mother would never have allowed anyone to approach my hair with a flaming taper. Indeed, as a child, I was never allowed to hold a box of matches, even safety matches, because, my mother was concerned that it might have spontaneously burst into flames. She should have known better. Her grandfather manufactured matches in South Africa.

It is easy to blame the mother for one's own fears and worries, as therapists are keen to suggest, but whatever the reason I have remained to this day fearful of any kind of medical, dental, optical, and cosmetic, interventions.

The fear of the nature of the intervention has acted against me several times. In 2014, I persisted in suffering uncomplainingly with a decayed molar tooth, which had an exposed dental pulp. I was terrified of undergoing treatment that would have got me out of trouble. So, it was only after several months that the ever-increasing pain drove me to seek relief from one of my colleagues. I must admit that once it was done, I realised that my pre-operative anxiety had been excessive and unnecessary.

The fear of undergoing medical procedures leads to many men, including me, to delay seeking medical attention. So it was in mid-September this year. One Tuesday morning, I found that although I had an urge to urinate, no more than a few pathetic drips came forth. At ever decreasing intervals, I was tormented by the urge to pass water. Each time, I visited the toilet, but flow was there none. That night, I was leaping out of bed every couple of hours, desperately feeling the urge to 'pee', but nothing was produced.

On the Wednesday and the two following days, I went into work. Luckily, none of my appointments were of long duration because in between each patient I made pointless visits to the toilet. As the week progressed, my nocturnal excursions to the toilet increased to such an extent that I was getting out of bed once an hour or even more frequently. I just hoped that whatever was causing my blockage was 'just a passing phase'.

After a miserable weekend, I made an emergency appointment at our GP (general practioners') practice. Without examining me, the arrogant GP, whom I saw, immediately diagnosed a urinary tract infection. I asked her whether it was possible that my problem might be related to my prostate, a commonly occurring problem in men of my age. "Definitely not," was her surprising response, "prostate problems never start so suddenly." Because I was so relieved to hear this, having heard 'horror stories' about faulty prostates, I did not pursue the matter further, and happily headed off to a pharmacy to get the antibiotics that she had prescribed.

Kindly, the GP rang me the next day to see how I was. I told her that I was neither better nor worse, and she told me, quite correctly, to contact the surgery if things got worse.

Tuesday, Wednesday, and Thursday passed, without me passing water. I cancelled work. I slept almost all day and night. I could hardly eat. I became snappy, uncivil. I kept taking the antibiotics religiously. Waves of pain began spreading up the flanks of my abdomen. My inability to pass water continued, as did the constant urge to do so.

On Friday morning, our daughter noticed that I could barely hold a glass of juice so bad was the tremor in my hands. She told me to go back to the GP. I told her that I would wait until my seven-day antibiotic course was over. However, she persisted, and I made an appointment for that afternoon. I am glad that she had nagged me to taking action.

At the GP surgery, I was seen by Dr A. In contrast to the first doctor I consulted, he was a model of clinical excellence. Being a clinician myself, I can distinguish between a competent clinician and a less competent one.

Dr A took a very thorough history of my condition, and then asked me to lie on his examination couch. Gently, he felt my abdomen, and then told me that my bladder was so full and over-extended that it was pressing against the underside of my diaphragm. He told me to go straight to a hospital. I chose the

Chelsea and Westminster ('C&W') because I had heard good reports about it. For at least twenty minutes, Dr A persisted on trying to establish contact with someone in the hospital's urology department. To his credit, he did not give up. Once he had spoken to someone, he printed out a letter to take to the Accident and Emergency ('A & E') section of the C&W.

I should have taken a taxi to the hospital, but because I could not think straight I took a bus. I could not walk straight. I must have looked like a drunkard as I staggered to the bus stop, and then from my destination stop to the hospital. The staggering was related to the muscle tremor that our daughter noticed. Despite my sorry state and discomfort, I was terrified of what was about to happen to me. The doctor had spoken about having a catheter inserted into my male organ, and that this would provide me with immediate relief. Yet, still I feared the worst.

After a short wait amongst the other A&E patients, I was ushered into a small room and lay on a bed. Straight ahead of me on the wall, I saw a long photographic frieze showing a flower-filled field. The colourful photograph was printed on a plastic material and lit from behind. The nurse pressed a button, and then the room filled with slightly sombre music, which immediately reminded me of the rooms in which the deceased are remembered just before their coffins roll silently towards the crematorium's oven. Not a great start, I felt.

The insertion of the urethral catheter was barely uncomfortable, and this minor discomfort lasted for a few seconds only. My bladder emptied. It contained four times what it should have done had all systems been healthy. Immediately, I felt marvellous. But, I should not have been so content because something, which I had not known about, had happened without me having the slightest knowledge about it.

I was sent up to a hospital ward where acutely ill patients were being cared for. Various cannulas were inserted into my veins, and I was attached to a variety of drips. I was monitored regularly for vital signs. I felt bemused, rather than uncomfortable. I was reluctant to move much lest the tubing attached to my catheter caused me pain in a very sensitive organ.

I was hungry. I had not eaten properly for several days, and I had missed the 6 pm patients' evening feeding time. At 10 pm, I was given a very basic cheese sandwich, which I wolfed down. Never before had a dull sandwich tasted so good. My wife and daughter went home, and I began to absorb the atmosphere in the ward.

A large Irish man opposite me, who was undergoing observation following a severe head injury, kept saying "Jesus Christ" and "Fuck it" loudly. Every now and then he would slip into his over coat and attempt to leave the ward. The nursing staff would then ask him where he was going, and he would reply: "Just for some fresh air." Then, they would encourage him to get back into bed. Even when they attached him to fluids via a cannula, he would undo the tubes before trying to escape yet again. I doubt that I got much sleep that night.

The following day, Saturday, I was visited by a team from the urology department. It was then that I learned the worst, what I had not anticipated, and therefore not been concerned about. The consultant, Mr D, told me that when I was admitted my kidneys were in such a poor condition that they were thinking of putting me on a dialysis machine. The urine, which I had not been able to expel, has filled my over-extended bladder and made its way back into my kidneys. This had caused them to fail, and there was a huge build-up in my blood of creatinine and other waste products that the kidneys are supposed to remove from my blood stream. This pollution of my blood had been the cause of the symptoms that had occurred between seeing the first GP and Dr A. I was in a very bad state, although I was unaware of it. I was told that the urine retention problem was most likely caused by prostatic enlargement, and that I would have to remain in hospital to monitor the initial phases of my kidneys' recovery, which the medical team hoped would occur.

During the middle of Sunday night, I was told that I needed to move to another ward. Rather neurotically, I thought that was because I felt that the head nurse did not like me. But, she told me that I ought to be pleased that I was moving because my condition was improving. It was no longer an emergency. I was wheeled on my bed through the huge empty atrium of the C&W. It was eerily quiet.

I reached my new ward, and my bed was parked next to a large window. The room was dark, and all that I could see were several elderly men lying still on their beds. Occasionally, one or other of them would groan or make some unpleasant sound. My suspicious mind told me that I had been brought to this ward to die alongside these morbid old men. I felt very sad, but a charming Polish nurse came to set me up for the night, and said very sweetly: "Welcome to Lord Wigran Ward. I hope that you will enjoy your stay with us." I fell asleep.

Next morning, I woke to find a small dark-hued man in orange pyjamas standing at the foot end of my bed. He occupied the bed beside mine. K muttered something, and then asked me to guess where he came from originally. I could not guess, but then – and I do not know why - he asked me whether I knew Afrikaans. I said that I did not, but my parents, who were brought up in South Africa, did. K told me that he came from Durban originally, but that his father had been Irish. He was, he informed me, a 'black Irishman.'

Then K did something that struck a chord. It reminded me of something that one of my now deceased relatives might well have done. K wandered to my bedside table and looked through the pile of my books on it. "A reader, heh? And you read classy literature." What K had done, without being invited, was exactly what my late uncle Felix would have done, also without asking. This little act made me feel at home in my new ward.

There were six bed spaces in the ward. Three on each side. K's bed was between mine and an Italian's. The latter had a deep, croaking, rather menacing voice. He had tubes sprouting out of all orifices, including his male organ. I never discovered what was wrong with him, but I enjoyed exchanging pleasantries with him in

Italian. His bed was by the door and opposite one containing a young man whose left leg was propped up. This poor fellow, a musician and an event-planner, had suffered a terrible fracture (while playing football), which required lengthy surgery. Every day that I was in the ward, he would be given 'nil by mouth' in anticipation of the complex surgery he needed. Every day at about 2 pm, he would be told that the operation had to be postponed because the surgeon had more urgent cases to treat. I admired his patient acceptance of the situation. However, on the last day of my stay, when he learnt that yet again he had been starved pointlessly all morning, he lost his cool, and hurled his mobile 'phone at a wall. The 'phone broke into many pieces one of which could not be found.

My bed was opposite that of N. Even when awake he seemed as if he was on death's doorstep. After a day or two, he perked up, especially after the man who visited the ward with a shop on a trolley slipped him some complimentary confection. N proved to be great companion, popular with the other patients and the nursing staff. A retired professional musician, he had played the saxophone in many leading West London venues. There was an empty bed separating him from the young man with the fracture.

K was mobile, and darted around our ward, making jokes and amusing us. He made frequent forays to the food storage cupboard, helping himself to outrageous quantities of food, which he secreted in his bed, under it, and beneath his pillow. The nursing staff became upset by the mess he was making.

When K was told that he was ready to go home, he made a horrendous fuss because, I believe, that he was lonely at home, and would miss the camaraderie in our ward. He received several visits from care workers, who had to work hard to persuade him to go home, not immediately as they wanted, but in three days' time. It was sad to see such a pleasant man demean himself the way poor K did. He left after I was discharged.

I had been supplied with much reading matter brought from home by my wife. I did not need to resort to this. There was so much to watch and hear from my bed that reading was unnecessary. Observing what was happening in Lord Wigran Ward was more enjoyable than good theatre. My fellow inmates were chatty. The ward was like a good club. I was not surprised that K did not want to leave.

Although lack of attention has never been a problem with me at home - I am spoilt for attention by my wife Lopa and our daughter –

this is not the case for every patient. Many of them only ever receive intense care and attention when they enter such a splendid hospital as the C&W. Even though they may be suffering, they receive pampering far in excess of what they can expect at home. Soon, they become used to this, and leaving the hospital eventually, and facing the realities of everyday life, may be a considerable emotional wrench for them. This was certainly the case for K.

I discovered that quickly I became dependent on having everything done for me, but what was done was only what was clinically necessary. Because I did not realise this, I soon forgot about daily necessities such as tooth brushing and routine hygiene. I was becoming so dependent that I waited, in vain, for the staff to do these mundane things for me. I was quickly becoming institutionalised.

Every morning, the ward was visited by a very cheery lady, who arranged our meals for the day, and served them. We were presented with a glossy multi-coloured menu with a bewildering choice of dishes that were described in such a way that they promised to be mouth-wateringly good. On the first day in the ward, I ordered an Asian meat dish for lunch and dinner. These proved to be disappointing. I learnt that the safest bet on the menu was to order a plain, but boring, sandwich.

The hospital's nursing staff included people from all around the world. There were several Italian nurses, all of whom spoke perfect English. One of them told me that the Italians, who were highly qualified, came to the UK because there was a great shortage of suitable jobs for them back in Italy. When I mentioned to another of the Italians that everything in the C&W was of top quality except the food, she advised me not to order the pasta!

Once a day, I received a visitation from the urology team. On my first day in Lord Wigran, a young urologist came to my bedside, but did not introduce himself immediately. I asked him if he was on Mr D's team. Immediately, he replied: "That bastard…" I was shaken by this vehement comment. Smiling, he put his hand into his pocket, and withdrew his identity card, which revealed that he was Mr D's son! Young Dr D drew a beautiful diagram and carefully explained what the team believed had happened to me.

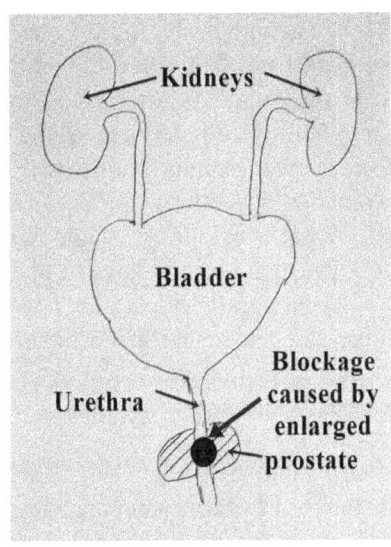

Simplified version of Dr D's diagram showing how the prostate can block urine flow from the bladder

Dr D told me that I would eventually need an operation to remove the blockage to my urinary tract caused by the prostate. This, he explained, involved no skin incisions because it was all done robotically via my urethra – if you don't know what this is, look it up! However, because my kidney function had been so badly damaged by the build-up of urine in my bladder, it would not be safe to give me general anaesthesia until (and, more worryingly, if) my kidney function improved. The reason that they were keeping me in the hospital was to do whatever was necessary to encourage the kidneys to recover. The doctor told me that when things began to improve in my renal 'department', I would be sent home with my catheter, and would have to wait at least two months before the hospital would consider scheduling an op. At some stage before that, my kidneys would have to be 'scanned', and then I would have to undergo an MRI scan. Although the young Dr D was most informative and reassuring, I did not relish the prospect of having the scans, and even less the general anaesthesia.

I had often visited sick relatives and friends in hospitals before I became someone who needed visitors to cheer me up in a ward. During my visits to my unfortunate friends and relatives, I always shuddered when I saw tubes coming out of their arms, bladders, and other places. I hoped that this would never happen to me. I believed that I would not have been able to cope with any of this psychologically. Yet, here I was in the C&W, with indwelling cannulas sprouting out of my arms, bottles of medicine dripping slowly into me, and a catheter directing my urine from my bladder to an enormous receptacle hanging by the side of my bed, making ill-considered movements of my legs extremely inadvisable. What I had feared greatly had, at last, happened to me. Surprisingly, it did not bother me at all. This was not because I had suddenly become transformed from a neurotic 'wuss' to a devil-may-care, water off a duck's back, 'cool dude'. The reason that I could tolerate what I had always feared was entirely due to the hospital staff. From the initial insertion of the urethral catheter onwards, they performed potentially unpleasant procedures with great care, gently, and often without causing any discomfort. I was very impressed by this. After a short time, anyone could take blood, insert a new cannula, or perform an examination, without my becoming even remotely concerned about it. This says a great deal for the skills and care exhibited by all who dealt with me in the hospital.

Day by day, my kidneys began recovering. Every day, the poor fellow with the fractured leg spent an entire morning waiting without food nor drink for an operation that was always cancelled at the last moment. I admired his patience. The delay had caused him to cancel a trip to see his fiancée in Russia, a trip that required him to obtain a visa not without a great deal of difficulty. He received many visitors including an attractive woman with Central Asian features, who gave him a massage. He spent hours on his 'phone talking in a respectably low voice. This irritated the Italian fellow in the bed opposite him. When the

Italian complained loudly, the young man told him that he was perfectly entitled to talk on the 'phone, and that there was no reason why he should not do so. This further annoyed the Italian, who became quite threatening. I was worried that a fight might begin, but then realised that, firstly although the young man was quite agile with his crutches, it might have been difficult to both fight and remain standing, and, secondly, the Italian was almost immobile because of the great number of tubes sprouting from his body. A nurse calmed the two men eventually.

N, who was in the bed opposite me, livened up gradually, and joined in with the largely jocular conversations that lightened the atmosphere of our ward. K, who should have been sent home, kept amusing us and upsetting the staff by pinching snacks from the ward's kitchen. It struck me that he had begun eating compulsively, as if he felt that there would be no tomorrow. A great collection of crumbs and debris collected around his bed. The floor cleaner became annoyed and reported his poor behaviour to the nursing staff, who told him off. He responded by laughing at them. My respect for K, which had originally been high, began to decrease.

When I entered the hospital, I had become unsteady on my feet. During my stay, this became worse. I had to be escorted to the bathroom, and had to use a Zimmer frame. After several days, I received a visit from a hospital physiotherapist. Her assistance helped me get 'back on my feet'. She gave me confidence rather than any other treatment. A day before I was discharged, I had to demonstrate to her that I could climb fifty-nine steps, that being the number of steps leading from street level to our flat. I succeeded. This was a sign that the toxin levels in my blood were falling.

Measuring the amounts of toxins in my bloodstream was an accurate way of assessing the degree to which my kidneys were improving. At regular intervals during the day and night, nurses and phlebotomists inserted butterfly wing cannulas into veins in my arm, and drew samples of blood for analysis. Everyone who attempted this was very good at it, but one person stood out from the 'crowd'. He first appeared in the middle of one night whilst I was asleep. I awoke to discover a thin man, whose appearance reminded me of elderly elves drawn by the illustrator Arthur Rackham, shining a torch at me. In a calming voice, he explained that he needed to take some blood. He reassured me that not only did he love his job, but he was also very good at it. He told me that he enjoyed a challenge. The harder it was to find a vein, the greater his feeling of achievement when he found it. His ambition was to become an operating theatre assistant. He lived up to his claims, and proved to be the best of all the people who took blood from me. His reputation was high amongst the ward staff. If he fulfils his ambition, the hospital will lose a top-class phlebotomist.

One day, it was deemed necessary to check my heart's functioning with an electro-cardiogram ('ECG'). About twelve electrodes must be attached to various points on the trunk and limbs. These are held in place by sticky plasters. Twelve wires sprout out of the ECG machine, and each one needs to be placed on the correct plaster before the ECG machine will begin to give a read-out. The machine can detect if a lead has been connected to the wrong plaster. A young doctor - she had been qualified less than one year - decided that she would wire me up. She was watched by two fully qualified senior nurses, who remained silent and expressionless as she fumbled trying to get each wire onto its correct plaster. Time after time she got it wrong. I wondered how long it would take because there are an awful lot of ways of connecting twelve wires to twelve different plasters. What made the whole attempt seem so silly was that the wires were of differing lengths and helpfully labelled, but Dr L knew best! When the gadget was finally wired up correctly, it produced a scan that revealed that there was some irregularity in my heart's electrical activity, and this was due from either too much or not enough Calcium in my blood. As Dr L was such a keen youngster, I dreaded her referring me to the Cardiology Department, but she did not. Instead, I was put on yet another drip, and given some drug as an inhalation. Next morning, one of the nurses, who had watched Dr L struggling, wired me up to the ECG within seconds, and, after looking at the machine's print-out, pronounced that my heart was behaving normally again. After all the wonderful care that I received at C&W, it is probably churlish to mention that the recently qualified Dr L, who demonstrated a certain amount of arrogance when dealing with my medication dosing, needed to work on her bedside manner. I told her that, and she took it remarkably well.

On my penultimate day in hospital, I was sent to the ultrasound department for my kidneys to be scanned. Contrary to what I feared this was a painless, and almost pleasant experience. A young doctor sat close beside me on a treatment couch gently massaging the flanks of my abdomen with the scanning device before reporting that she could not detect any structural damage to my kidneys. This was reassuring news.

The day of my discharge arrived. By 2 pm, I received my discharge letter, some medications, and some bags and other equipment for managing my in-dwelling urethral catheter. Apart from telling me which item did what, I received no further information on how I was to deal with life wearing a catheter. I sat in the 'day room' fully dressed and with a urine collection bag strapped to my leg. Walking was very difficult. Lopa collected me, and we went home in a cab.

Despite my great and real fears of ever having to undergo even one of the procedures that I experienced during my six days at C&W, I must admit that my anxieties were unfounded. I had not been a hospital in-patient since 1962, when I had to have my appendix excised surgically. My memories of this are fond.

I was admitted with my 'grumbling' appendix to the private St John and Elizabeth Hospital in St Johns Wood, close to the Lord's Cricket Ground in north-west London. This hospital was run by Roman Catholic nuns. I was given a private room with a television ('TV'). We did not have television at home (and still do not!), so to have my own TV to myself was a real treat. I watched all day, and used to get annoyed when visitors kindly came to see me, because then the TV had to be switched off. I recall little of the operation except for two things.

The first was just before being taken to 'theatre', I was taken to a bathroom, and told to have a bath or shower. When I was finished, I was supposed to pull a cord to summon a nurse to escort me back to my room. I washed, and then pulled one of several cords that were dangling from the ceiling. Nothing happened, no one came. I peeked through the door that led from the bathroom to the corridor. I watched with amazement the chaotic activity in the passage way. Nurses were running about, trolleys were being wheeled hither and thither. Someone came to me, and asked me whether I had pulled the fire alarm cord. I must have done, but out of ignorance because no one had told me which of the various cords I was supposed to pull.

The other thing I recall was after the operation when I spent several days in hospital. Contrary to today's thinking, patients who had just been operated were required to move as little as possible. Nowadays, barely have you recovered from the anaesthetic and you are encouraged to 'mobilise', to use current hospital jargon.

Anyway, back in 1962, I was told to remain in bed. If I wanted anything, I was to ring a bell by my bed. It was activated by a button attached to a plaited, cloth-covered, electrical flex, which would be frowned on by today's health and safety people.

TVs did not have remote control systems for changing channels. In the UK, there were only two channels available in 1962: BBC and ITV. What I needed most often whilst lying in bed was to change the TV channel I was watching. I would ring the bell, and a nun would come rushing in. Uncomplainingly, she would turn the knob on 'my' TV so that my desired channel could be viewed.

My appendix operation was a long time ago. I was ten years old, so I might have forgotten several things about it. However, I have no recollection of being 'wired' up to drips or having long-term cannulas in my veins. I suspect that my medical condition had not been serious enough to require this kind of thing.

Returning to 2016, I went home, having been told that until my kidneys had recovered as much as possible no surgical procedures would be contemplated. I was to have regular blood tests, and eventually I would be summoned for surgery. I resumed daily life with my indwelling bladder catheter. This tracked from my bladder through the prostate and projected from my male organ. Urine could flow through it unimpeded by my prostate and without my body being able to control it. It had to go somewhere, and I was given various attachments to collect it. One of these was the so-called 'leg bag'. This was attached to a long tube that had a plastic end that could be inserted into the part of the catheter projecting from my body. The half-litre capacity bag was strapped to my leg below the knee by two elasticated bands. When it filled, which it did quite quickly, I could empty it into a toilet via a tap built into its base. In bed or at home, the leg bag did the trick, but in the world outside our home, it was a great impediment. I could walk with the bag attached, but only slowly and awkwardly. My gait became a sad-looking limp. I dragged myself along the street, and must have looked like a crippled, old man.

After a few weeks, I decided to try another gadget which had been supplied to me. I was reluctant to use it at first because the urologists were worried that the urine retention and the consequent gross stretching of my bladder might have damaged the organ's musculature to such an extent that in might not have been able to expel urine in the future. The other gadget was a small tap (a 'spigot'), which I had to insert into the external orifice of my catheter. When my body told me that the bladder was filling up, I could simply go to a toilet and open the tap.

The spigot was far more convenient than the bag. I knew when I needed to open it because I could feel stirrings in my lower abdomen. These were not like the usual sense of urgency that I normally feel when I need to urinate, but they gave an accurate indication of when it was time to undo the tap. With the spigot, walking became a bit easier, but remained somewhat abnormal. The limping gait was inevitable, because I had to take care not to make movements that would cause the catheter to tug on my 'member', and cause it to become painful. On the whole, I felt well and in good spirits, but I found that I was tiring more quickly than before my kidneys began failing.

The urologists had told me that I would have to undergo a Magnetic Resonance Imaging scan ('MRI' scan) to check the size and shape of my enlarged prostate.

I first heard of magnetic resonance whilst studying biological chemistry as part of my physiology degree course at University College London. Nuclear magnetic resonance spectroscopy is used to investigate the physical and chemical properties of molecules, and is of great usefulness to organic chemists. On the other hand, medical MRI scanning allows a non-invasive investigation of body parts (including soft tissues) without any dangers such as harmful radiation.

Many people who have experienced MRI scanning have told me how fearful an ordeal it is. Their main concern was having to lie still for a long period in a noisy, featureless, confined space in a narrow tube that is barely large enough to hold a body. When I learnt that I was going to have undergo an MRI scan, I was filled with anxiety. Even though I had already experienced far worse - and without finding it at all unpleasant - I was not looking forward to having my scan.

I arrived at the scan, feeling like the peanut which stood on the railway track, whose heart was all a flutter (when 'around the track the engine came… toot toot peanut butter').

Putting a brave face on it, I entered the scanning room through a reinforced metal door that looked like the entrance to an atomic bunker. I lay on a narrow bed, which turned out to be extremely comfortable. Before being given a set of headphones to protect my ears from the noise that would be produced during the scan, I was asked what kind of music I would like to hear. I asked what was on offer. The choice was between Motown and classical. I opted for the latter.

The bed, with me on it, slid slowly into the circular tunnel in the centre of the Siemens 'Magnetron'. I continued entering it until only the crown of my head was outside it. When I looked up, all I could see was the grey funnel-like rim of the entrance to the machine.

There was a sound like a fog horn, and then the sound of monotonous soporific classical piano music, rather tinny in tone, came through my headphones. No decent composer would have had the gall to own up composing this pathetic attempt at 'classical music'. Nevertheless, it was mildly distracting and its lack of variety helped me to relax.

Then, the fun began. For reasons that the nurse could not explain the MRI machine produces a series of extraordinary noises, which must have been very loud because I could hear them quite clearly despite wearing the ear-protecting headphones. The first of these noises resembled someone hammering loudly at a building site. This was followed by bursts of sound (each lasting several minutes) that included 'kerchunk, kerchunk, kerchunk,…'; 'boop,

boop, boop…'; 'whooo, whooo, whooo,…', 'tak,tak, tak…'; and so on. All the time, the monotonous piano music droned on, barely competing with the miscellany of bursts of weird mechanical sounds coming from the magnets in whose womb I was confined. At several stages, the machine seemed to become over excited, not only emitting noises but also causing the bed on which I was lying to vibrate.

Far from hating the whole experience, as I was sure that I would, I found it mildly entertaining. The forty odd minutes of my scan shot by. Let me explain. First, I was extremely comfortable. Having to lie still on a comfortable bed was very restful and relaxing. It was far more comfortable than sitting for forty minutes in an aeroplane or in some theatres. Second, the noises conjured up various images in my mind. During the vibrations described above, I felt as if I was on one of those reclining chairs in Business Class on a long-distance flight. The odd combination of the repetitive classical music accompanied by the series of ever-changing mechanical noises being emitted by the scanner resembled the music of minimalist composers, notably the compositions of Steve Reich. At times, I felt as if I was listening to a bad pianist giving a concert in a busy construction site.

Many years ago, I attended a concert of Spanish Flamenco dancing at the Victoria and Albert Museum. The endless racket produced by the dancers stamping their shoes on a hard floor was far less bearable than what I heard during my MRI. During the scan and after, I felt that the horror stories, which I had heard about MRI scans, should possibly be discounted.

Several weeks passed. I saw friends, finished writing a book and published it, went to the theatre, and visited my GP practice for routine blood tests. Each successive blood test result revealed that gradually my kidneys were recovering. Eventually, I was summoned to the C&W for 'pre-operative assessment'. As with everything at this hospital, it was beautifully organised.

On arrival at the pre-op assessment department, I was asked to fill in a personal health questionnaire, and then invited to wait on one of the comfortable armchairs supplied nearby. Earlier than my appointment time, a young oriental lady invited me into a small room where she measured my blood pressure before efficiently performing a full ECG at lightning speed. Then, I returned to the waiting area.

At exactly my appointment time, an anaesthetic nurse took me into another room, and rechecked my medical history with impressive thoroughness. She explained various things about preparing to have general anaesthesia ('GA'), and what I was to do on the day of my operation. Having explained everything, she sent thorough detailed instructions to my email address. I asked her several questions including when my operation was due to be carried out. She told me that it had not yet been scheduled, but the pre-op assessment remained 'valid' for three months. Then, she sent me to have some blood samples taken in the out-patient department that deals exclusively with that.

The blood-taking department was a model of efficiency. On entering, patients take a numbered slip of paper from a dispenser, and then a seat in the waiting area. When my number was called, I was directed into a small room that resembled a hairdressing salon. There were three chairs in a row. I was directed to one of these, and before I had time to think, a very well organised oriental lady, began taking the necessary blood samples, painlessly. Within less than a minute, she was applying a plaster to the tiny wound in my arm. I asked the phlebotomist how many patients she dealt with each day. She said: "Hundreds".

Having completed the pre-operative assessment procedures, I returned home, and awaited the arrival of a communication from the hospital regarding the date for my operation. After two weeks, I had heard nothing, and imagined that there might be a long waiting list for surgery – this is not unusual for the NHS. I waited another couple of days, and then rang a number I had been given for the surgical appointments department. To my great surprise, the person who answered me said: "When would you like to come in? You can choose the day." I could not believe that I was being offered a surgical appointment more easily that I could book a table at a restaurant. I chose a convenient date about nine days ahead, and then began to worry.

I had been told that the surgery was relatively minor. It was not that which concerned me. It was the general anaesthesia ('GA') that caused me to have great anxiety. The rational part of my mind told me that GA these days is incredibly safe; little could go wrong. After all, my frail ninety-year old mother-in-law survived anaesthesia supplied during a far more complicated procedure than I was about to undergo. But what if it did go wrong? What if I was to become the few, who did not come 'back to life' after GA? According to an NHS advice sheet, the risk of death during GA is between one in 100,000 to one in 200,000. Even those figures worried me. By deciding on an operation date, I felt that I had risked determining the date on which my life was going to come to an end. I kept telling myself that I was being ridiculous in worrying about a pessimistic outcome, but my anxiety was never far from the forefront of my mind.

Imagining that I had only just over a week to live, I decided to fill my time as fully as possible. In five days, I saw two films, attended a classical music concert, enjoyed a work's Christmas party, and watched our daughter lecturing about a painting in the Courtauld Gallery. I read as much as possible, and finished publishing a book that I was working on, as well as editing the proofs of another one. I am sure that I must have upset my wife by endlessly voicing my anxieties about not recovering from GA. She was already worried about my having surgery, but not about the GA. She had survived many episodes of GA.

The day before the date of the surgery was awful for me.

My fear of not recovering from GA had got me to the state of wondering whether I was doing each mundane thing for the last time. Was I having the last shower in my life? Would I never again adjust the thermostat on the central heating? At breakfast, was I eating my last piece of toast? Would I ever finish the novel that I was reading? It was "Colomba" by Prosper Mérimée. Was I taking my last ever ride on a bus? Was I saying "good night" to our daughter for the last time? That is how bad my fear and pessimism had become.

I awoke at five in the morning of the day of the operation. This early awakening was necessary because I had been instructed to drink as much as possible up to two hours before the appointment time, which was eight in the morning. We arrived at the hospital at about half past seven. Soon after registering my arrival with a nurse, I was called by an attendant. I had barely any time to embrace my wife – possibly for the last time – before I was whisked into the men only changing area.

I was instructed to get dressed in a couple of hospital gowns, and to put a couple of elasticated socks, one on each leg. These socks, which are fiendishly difficult to put on, are designed to help reduce the risk of deep vein thrombosis ('DVT') during the operation and after. I noticed that they were designed at University College London, where I had studied for twelve years.

Once I had struggled into the socks and locked my clothes into a numbered locker, I pinned my numbered key to my gown and took a seat amongst the few other male patients in the waiting area. There was a man sitting there who looked remarkably familiar. After a while, I told him that I recognised him. He acknowledged this grudgingly. I remembered that he had once been one of my dental patients, an incredibly grumpy fellow who was never satisfied with whatever I tried to do for him. Some years ago, I named a villain in my novel "Aliwal" after him. True to his old form, he was soon pestering the nursing staff about something or other. My other 'companions' were pleasant enough.

After a short while, I was taken into a small room to have my blood pressure measured and to be weighed. Also, I was asked to sign a disclaimer form absolving the hospital from any blame if my possessions went astray. Then, I sat in the waiting area for at least an hour before I was taken to have a discussion with the surgeon. He explained the procedure (transurethral resection of the prostate: TURPS, for short) to me before asking me to sign a consent form. And then, it was back to the waiting area for another long wait. By now, I knew that my operation would be the third to be done that morning. Not much was likely to happen to me until eleven o'clock at the very earliest.

At about eleven, I was summoned to speak with the anaesthetist. The first thing he told me after asking me when I last ate and drank was that I was likely to be conscious during my operation. It was considered to be far safer to do my operation under spinal anaesthesia (epidural) rather than under GA. The reason is as follows. During the prostate procedure, which I was about to undergo, much fluid needs to be flushed through the bladder. This could easily cause dangerous fluctuations in the blood's electrolyte (mineral) balance. Under GA, these life-threatening changes are exceedingly difficult to detect, whereas if the patient is conscious, the signs of electrolyte imbalance become easily obvious to the anaesthetist.

I was greatly relieved to learn that I would not have to undergo what I was dreading most: GA. I told the anaesthetist that, and he expressed surprise that I had not been told this weeks before when I had the pre-operative assessment.

The anaesthetist did warn me that it might become necessary to put me under GA despite the spinal block. I asked him under what conditions he would do this. He told me that some patients 'freaked out' (my words, not his) in the clinical environment of an operating theatre. In such circumstances, GA might become needed. He said that I seemed sensible, and likely to tolerate surroundings during surgery. I asked him the risks of side-effects from the spinal block. He told me that one in a hundred people get a terrific headache, and a very few people suffered permanent paralysis of the lower limbs. The risk of the latter was, he said, far smaller than the risk of being run over whilst crossing a road.

That reminded me of something a dentist once told me when there was a huge fear about the toxicity of silver/mercury amalgam fillings. He said that the risk of coming to grief with these was so small that he would happily replace the allegedly life-threatening fillings when the patients promised to stop smoking or crossing the road.

I returned to the waiting area with a spring in my step. It seemed unlikely that what I had been worrying myself sick about was going to happen. I had been given a reprieve from GA. Ok, well there was the tiny risk of paralysis and life in a wheel-chair, but my life was likely to continue one way or another after the operation.

I sat waiting. At about midday, a nurse apologised to me for the delay, saying that some equipment needed for my operation had to be collected from 'upstairs'. For a few minutes, I began becoming concerned that maybe the missing instruments would not be found, and then all my waiting would have been in vain.

Thirty minutes elapsed before I was summoned to the operating theatre. Before that, I had rung my wife to let her know that it would be quite a while before I was going to be 'sorted out'.

In the theatre, where soul music was being played through a sound system, I was given a light sedative intravenously. I began to feel as if I was floating. I was asked to sit on the edge of the operating table, leaning my head forward so that my back was extended. A small local anaesthetic was administered – I barely felt it. Then, the anaesthetist had to find the correct point at which to administer the block injection. At one point a brief wave of what felt a bit like a mild electric shock spread down my left leg. When the operator learnt of this, he adjusted the positioning of his needle. Within seconds, my legs began feeling heavy. I was just about able to swing them back onto the table. Soon, my bottom became numb. I felt as if I had lost the bottom half of my body. A crude curtain made from a blanket was rigged up using surgical tape in order that I would not be able to see what the surgeons were doing.

I laid back, chatted with the anaesthetist and his assistant, listened to the music, and watched the waves of activity on a bank of monitors close to me. I felt something massaging my legs. This was a pair of cuffs that inflated and deflated rhythmically in order to prevent DVT from beginning in my outstretched legs. Otherwise, apart from a little gentle tugging, I felt nothing during the surgery. I could not believe how un-stressful an experience I was undergoing.

When the surgical procedure was over, I mentioned to the surgeon, that the whole thing had been quite pleasant. He looked at me, astonished, and said: "Nobody has ever before described being operated as being pleasant." Well, truth be told, the whole theatrical performance had been anything but unpleasant.

I was wheeled into the recovery area, where there were a couple of patients, wearing oxygen masks whilst slowly regaining consciousness. In comparison with them, I was lively. I was extremely hungry. It was after half past one. I had not eaten for about sixteen hours. One of the very friendly nurses brought me a cup of tea and a good selection of biscuits. I was delayed a long time in the recovery room, because the ward where I was to be taken was not ready for me. The nurses kept ringing the ward, and returning to me apologetically. But, I was not worried. I was very comfortable, and greatly relieved that at last the surgery had been done.

Eventually, I was wheeled up to David Evans Ward. Oddly, the C&W does not have dedicated trolley lifts, so I, lying on my trolley, and my nursing escorts, had to wait alongside members of the public and hospital staff queuing for a lift.

Like the Lord Wigran Ward, where I had spent a few days recovering from the damage to my kidneys, the David Evans was divided into rooms each with six patient beds. This time, I was not beside an outside window, but by the corridor, separated from it by a wall with a large window. Following the surgery, I felt no pain. A urethral catheter, wider than the one I had before, was in place. This one was designed so that fluid (saline) could be flushed through my bladder from large three-litre bags (packed in Switzerland, of all places) suspended above me. The purpose of this was to wash out debris from the surgery, and blood. I was told that until what was being washed from my bladder was the colour of clear, pale Rosé wine, the flushing would need to continue. At first, the colour of the effluent was closer to Claret or cloudy Burgundy wine. It would take about one or two days before this transformation would occur.

The saline was whooshing through me so fast that the poor nurses had to replace the three-litre bags almost every twenty minutes, day and night. At slightly greater intervals, one of the charming team of nursing staff came to check my blood sugar, blood oxygen levels, temperature, and blood pressure. The oxygenation of the blood is measured using a sensor that clips onto a finger like a clothes peg. I guess I must be a rather lazy breather because whenever this measurement was done, the nurse would tell be to breath more deeply in order to achieve an 'acceptable' oxygenation level: 95 was good, anything lower bad!

Apart from slight irritation from the catheter, to which I was accustomed (having had one in place for two months already), I was not in pain. At intervals I was offered paracetamol, but did not need it.

The other men in my little ward were not as much fun as those with whom I had shared a ward two months earlier. One was a Labour councillor, who watched television much of the time. I tried engaging him in conversation, but he was reluctant. Apart from asking my wife and I whether we supported the Labour Party, he was not particularly chatty. Maybe, he was suffering pain.

Next to me, the bed was empty the first night. The next night, the space was filled by a Scotsman who had undergone major abdominal surgery. He was amicable, despite the great pain he was trying to endure. Whenever the nurse asked him to rate the level of his pain from 0 (no pain) to 10, he would say "8" or "9", where I never rated my discomfort as more than "1".

The man in the bed opposite me, T, kept out of sight most of the time, and I could understand why. One side of his face was partially covered with often bloody dressings. The exposed part of his face was often blood-stained. He told me after a while that he had just had a skin graft to correct some facial asymmetry that had occurred many years ago, because of surgery for a carcinoma. I felt sorry for him, but did not find him much fun. Although he was suffering a great deal, he made sure that no one was unaware of it.

At odd intervals, day and night, T would begin shouting almost hysterically. He yelled things such as: "Help, it's come off!", "Oh, no I've lost it in my bed", and "I don't like this. It's causing me so much stress." These cries attracted the nursing staff, who entered his curtained off area, and then sorted out the problem, after which he always said repeatedly: "Thank you so, so much".

The nurses kept entering T's 'territory' carrying small plastic pots rather like those that contain the hummus (the Middle-Eastern chickpea based spread) that is sold in supermarkets. The pots carried to and from T's bed appeared to contain crushed ice, but, as I soon found out, that was not all. The other ingredient in the pots was a longish black slimy object, and this was what was causing T so much worry.

T was having leeches applied to the site where his skin graft had been attached to his face. I was astonished because I believed that the use of medicinal leeches had 'gone out with the ark'. Nowadays, these slimy creatures have become the surgeon's little helpers. When the leech 'hooks' itself onto a victim, it secretes proteins that prevent blood from clotting. The surgeon wants to prevent blood from clotting around the graft reception area because this somehow promotes the migration of small blood vessels from the reception site into the graft, which needs to be kept alive if it is going to take. The problem is that the patient has to endure excessive bleeding. This was one of the many things that T did not like about having leeches applied.

The well-behaved leech is supposed to cling on to the graft area until its little body becomes swollen with blood, and then it is supposed to fall off or be removed. The trouble with T was that his leeches often fell off before they had done their job, frequently a few minutes after they had been applied. I have no idea why they did not sit where they were supposed to for a while, and then enjoy a good meal. My theory was, but I did not reveal it to T who was in no mood for humour, that the leeches sensed that they were not welcome. They sensed that T was hostile to them, and that is why they leapt off as soon as possible. Well, what do I know about leeches?

A used leech cannot be recycled, even if it has only sat on a patient for a few minutes. The 'retired' leech has to be killed by squirting it with alcoholic hand cleaning gel. Incidentally, all over the C&W hospital there are dispensers of this 'leech-icidal' gel. In the public areas, a recorded announcement is made repeatedly, exhorting all and sundry to make use of this gel in order to prevent spreading infection to patients.

One evening, T became particularly nervous. He learnt that the night sister had had little or no experience dealing with leeches. Would she be able to cope with them? Well, his anxiety was not groundless. The leeches were even more capricious than before. These expensive creatures were dropping off his face almost as soon as they reached it. The panicky sounds coming from T were endless. At one point, a sister carried one of the small containers away from T's bed, holding it and the black leech, which it contained, at arm's length with an expression of fear and disgust on her face.

Next morning, another nurse, who was good with leeches, told me that one of the used leeches had escaped. The night nurse, who was supposed to have killed the leech, must have been an animal lover, because she had not sprayed the now fugitive leech with the hand gel. She had thought it was cruel to 'murder' the poor little leech. Despite the scares and alarums, the leeches were actually doing some good, and T was pleased with what the surgeon had managed to achieve.

Meanwhile, I was in bed, quite comfortable whilst uncountable litres of saline were being pumped through my bladder. Eventually, the Claret/Burgundy effluent from my bladder became Rosé. I was visited by the urologists once a day. They appeared pleased with how things were going.

Refreshments and meals were served by lady from Goa. She spoke Hindi amongst other languages. After a brief misunderstanding between her and my Indian wife, the two ladies became good friends. Because of my Indian 'connection', the Goan made sure that I had the best service she could supply. If Lopa arrived late at visiting times, she would come up to me and ask concernedly why my spouse had not yet arrived.

On the third day of my stay, the clinicians deemed that the flushing had achieved what was required. It was time to remove the catheter. I was not looking forward to this, expecting it to be greatly uncomfortable, and, also, wondering whether I would have become incontinent following the surgery. Nothing happened for hours, apart from routine glucose testing, blood pressure measurements, and so on. Lunch was served.

I had spotted the Halal dishes on the menu that the Goan lady brought around in the morning. I ordered Chicken Korma, and the Goan asked me whether I liked my food spicy. What arrived at lunchtime was better than anything that I had eaten either on this or my previous stay in C&W. It was good, far more tasty than a supermarket 'ready meal'. I would have been happy to serve this to discerning dinner party guests.

A couple of hours later, a nursing sister arrived to remove the catheter. She must have noticed me tensing up, and reassured me that the removal would be quick and only mildly uncomfortable. I knew that the catheter was prevented from slipping out of the bladder by a balloon that was inflated once the tube had been placed within it. I had thought it was inflated with air, but I was wrong: it was water. After deflating the balloon by withdrawing several syringes-full of water from it, the nurse eased the catheter out of me. It was, as she had said, mildly painful, but it was quick.

To my great relief, I was not incontinent. Nothing leaked from where the catheter had been. When I felt the urge, I visited the toilet. I could not believe the force and speed of my urination. I was reminded of the sudden, almost miraculous, speed at which our kitchen sink, which had been blocked, drained after a visit from the Dyno-Rod plumber, who had used special mechanical cleaning machine to unblock our pipes. With apologies to the surgeons, who have infinitely more skill and knowledge than the plumber, I must admit that I felt as if my urinary tract had been not only unblocked but, also, improved.

Several hours after the catheter was removed and the nursing staff were happy with my ability to 'pass water', the nursing sister eventually managed to persuade a doctor to write a discharge summary for me and my GP. Then, I returned home to begin my recovery.

I have written this somewhat graphic account in order to reassure the many men, who like myself, live in fear of having prostate problems. It has struck me that during my period of illness, everyone has taken an interest in my well-being, but most of all the men of a certain age, with whom I have spoken, have taken a greater than usual interest in what I had to say. If I had said that I was suffering from the influenza or diabetes, most of my male friends would have rapidly moved on to another topic of conversation. When I mentioned 'prostate', their interest became intense. They wanted to know as much as possible. One man listened to my tale of woe for about ten minutes, before saying: "There is nothing I have feared more than having trouble with my prostate. Having spoken with you, Adam, I feel greatly reassured and less worried."

I hope that whoever has read this book will feel that his concerns have been reduced somewhat. I hope that female readers, who have plenty of their own potential gender-related medical problems to anticipate, will have found my story of some interest.

Finally, I wish to express my sincerest gratitude to everyone at C&W Hospital, who made me comfortable, healthier, and calmer. I am grateful that they made even the most difficult things seem so simple and easy to tolerate.

I would like to thank my wife Lopa for her assistance with literary aspects of this work, and for her affectionate support during the awkward moments of my illness.

www.ingramcontent.com/pod-product-compliance
Lightning Source LLC
Chambersburg PA
CBHW072252170526
45158CB00003BA/1057

* 9 7 8 1 3 2 6 8 9 4 9 4 8 *